150 TIPS FOR **BETTER GAS MILEAGE**

Over 150 Things You Can Do to GET MAXIMUM MILEAGE Out of Every Gallon of Gas

LUKE MELTON

A "Reality Check Guide" to Living Well in the New Normal

www.RealityCheckPublications.com

DISCLAIMER AND CREDITS

This is a common sense guide to getting better gas mileage from your car. The information contained herein is based on the best information available to the author at the time the booklet was written and is provided in good faith that the techniques discussed have every reasonable chance of improving your fuel economy to some extent. Due to the many unique and individual variables associated with how you maintain and drive your car, the author makes no guarantees, explicit, implicit or implied, regarding how much your fuel economy will actually improve or how the techniques detailed herein will affect your vehicle. Readers are cautioned to assess the information provided and to rely on their own judgment concerning how to use this information in their own individual circumstances.

This book was designed, written and illustrated by

Luke Melton

at Latitude 28 Design and Marketing
2174 Greenbriar Boulevard, Clearwater, FL 33763 USA
Website: http://www.latitude28dm.com
Email: luke@latitude28dm.com
727•738•9264

For more information about this book, or to purchase a digital download or a printed hardcopy,

please contact:
Reality Check Publishing
2174 Greenbriar Boulevard, Clearwater, FL 33763 USA
Website: http://www.tipsforbettergasmileage.com
Email: sales@realitycheckpublishing.com
727•738•9264

Published May, 2012 by
Reality Check Publishing

© Copyright 2011-12 by Luke Melton. All Rights Reserved.
No portion of this document may be copied or reproduced in any form without the expressed, written permission of the copyright holder.

Image of the Hyundai Genesis Coupe on the cover
of this book is provided by Hyundai Motor America.

TABLE OF CONTENTS

PAGE	SECTION TITLE
v	Foreword
1	Introduction
3	Take Care of Your Tires
5	Check Your Tire Tread Depth
8	Properly Maintain Your Car
9	Speed and Gas Mileage
12	Gas Mileage Techniques For the Road
14	Improving Gas Mileage While Driving In Town
17	Coasting Saves Fuel...But
17	Using Your Air Conditioning System
20	Use The Grade of Gas Recommended By Your Car's Manufacturer
21	Search Out Where to Buy the Least Expensive Gas In Your Area
25	Use the Internet and Your SmartPhone to Locate the Cheapest Gas
27	Keep Consistent Records of Your Fuel Economy
29	What About Commercial Gas-Saving Devices?
30	Find and Take Advantage of Your Transmission's "Sweet Spot"
33	Use Your Car's Fuel Economy Gauge to Verify Your Gas Mileage
34	"Plan" to Get Better Gas Mileage
37	Change Your Lifestyle To Get Better Gas Mileage
42	Buy a More Fuel-Efficient Car or One That Runs On Alternative Fuel
44	Hypermiling Techniques To Avoid
46	The 10 Most Fuel-Efficient Cars
47	Fuel Economy Links

The author would like to thank the following people, who contributed substantially to the production, publishing and marketing of this book:

Susan Kelley • Peter Scribner • Mike Gusler

Dan Sherman • Barbara Grassey

FOREWORD

We are living in challenging times. Many are calling the circumstances we now face "The New Normal", a state of affairs in which the rate of change we face is faster than ever before in human history. Among other realities in the New Normal of 2012, most of us are dealing with a particularly disturbing one: plummeting incomes combined with ever-increasing prices for nearly everything we buy.

This leaves many folks in a real bind: how to parcel out their incomes to pay the mortgage, buy food and clothing for the family, pay medical expenses and ever more pricey prescription medicines...and a big one, **pay for the fuel we need to feed our cars if we are forced to drive to and from work**.

Does it feel like you are increasingly working just for gas? Gas is one of the most obvious of the commodities for which we're paying more: the price of petroleum refined into gasoline may stabilize for a short while, but the overall trend is constantly upward. And most of us don't know what to do about it.

This book gives you real tools to do what is called *hypermiling*, making skillful changes in the way you drive to reduce the cost of using your car to get to work, go shopping, drive on vacation, etc. Here you'll find over 150 tips to coax a greater number of miles out of every gallon of gas you put in your gas tank.

Much of what you're about to read involves modifying your driving habits. But, having grown up in a time in which gas was relatively cheap and we could burn it with impunity, many of us will find it difficult to change the mind-set that governs how we drive our vehicles to conform to the constraints of the New Normal.

So, here's a program to make it easier to modify your driving habits and really improve your gas mileage:

- First, understand that implementing many gas-saving techniques requires a series of minor changes in your driving habits. But, **don't try to change everything at once**. There's literally no way you can incorporate all the tips in this book at one time. Instead, set a goal to select one or two tips that appeal most to you and spend the next few days building them into your driving mind-set.

    Write these tips down and tape the list to the bathroom mirror so you'll see it each morning as you brush your teeth. Make a separate list to keep in your car for reference. For example, the first week you might start to ease off on the accelerator when approaching a red light, giving the light time to change to green before you come to a complete stop. During week two, you might begin becoming more aware of what traffic is doing a light or two ahead.

Gradually, the tips you incorporate will become true habits, and in a few months you'll be driving more frugally, saving a great deal on gas.

- Many tips don't require a change of habit so much as simply taking time to *do* something. For example, you might promise yourself that this coming Saturday you'll remove that big roof rack from your car, pull out all the gear that's accumulated in your trunk and stow it in the garage, and, most importantly, check the air pressure in all four tires, including your spare. There...that wasn't so hard, and you can now check these tips off your list temporarily (don't forget to revisit them periodically, however, as you'll likely be tempted to stuff more gear in the trunk in the future, and you definitely should check your tire pressures at least once a week).

- Other tips fall into the category of "I hope I'll remember that when the time comes." An example is deciding to look at more fuel-efficient car models, or hybrids, or even cars that run on alternative fuels, when it comes time to trade in the old clunker and buy a different car. You don't need to make this a habit, but it would be good to remember the tip when you start shopping for a new ride.

We wish you great success in implementing the suggestions in this book, and hope you find the rewards to be well worth the effort. Be sure to help your friends save gas: tell others what you've learned, and, if you find the cost of this book has been more than rewarded by the benefits, suggest they buy a copy. We would appreciate it.

TIPS FOR GETTING BETTER GAS MILEAGE
— MORE THAN 150 THINGS YOU CAN DO TO SAVE GAS —

Honda Insight Hybrid

Before we discuss gas-saving tips, let's take a quick look at some of the physics that govern how much energy it takes to move your car, energy that is supplied by burning the gasoline you pump into your car's gas tank. Don't worry that this is "physics"; I know this isn't the most pleasant way to introduce a book, but this rather fundamental information is important for you to know, and, as you'll see, it's really quite understandable (on the other hand, feel free to skip this section if you *really* hate physics).

- First, it's important to understand that, depending on the total weight of your vehicle (GVW = Gross Vehicle Weight), **it takes a *given minimum amount of energy* to make the mass of your car *start moving in the first place*, a *different* amount of energy to *accelerate* the vehicle up to a specific speed, and yet a *different* amount of energy to *keep it moving at a given speed*.** This is the energy required with no other factors impeding the car's movement.

    A drop of gasoline provides a **specific amount of energy** when ignited in an internal combustion engine; in order to provide the energy required to move your car, a specific amount of gasoline must be burned. The rate at which gas is consumed to keep your vehicle moving at a specific speed for a specific period of time is the **absolute minimum amount of gas required to operate your car at a specified speed**. No matter what you do, **to travel a certain speed for a given period of time will always require your engine to burn a specific amount of gas for that period of time**. This is the ***very best gas mileage you will be able to achieve for your particular car traveling at that specific speed, and you cannot increase your mileage beyond this limit***.

    For instance, (here's some of the physics) let's say that it requires 1,000 foot-pounds (ft/lbs) of energy to move your 2004 mini-van (which weighs 3,675 pounds) along a level highway at 45 miles per hour for one hour.

Given the amount of energy a gallon of gas can generate, to deliver 1,000 ft/lbs of energy constantly for that hour your engine must burn 2.76 gallons of gasoline. This means that you will get 16.3 miles per gallon during the hour you're traveling (45 miles divided by 2.76 gallons = 16.3 mpg).

If you drive a larger, heavier SUV (with a GVW of 4,300 pounds) it may take 1,300 ft/lbs to keep it traveling at 45 mph. In this case, to generate 1,300 ft/lbs of energy constantly for one hour you'll burn 3.50 gallons of gasoline, and your gas mileage will drop to 12.86 mpg.

In both cases, the 2.76 gallons and 3.50 gallons respectively provide the minimum amount of energy required to keep the two vehicles moving at 45 mph. This is the **absolute minimum amount of fuel that must be burned, and you cannot improve on this mileage** (the mileage can, however, *decrease* as the result of other factors, such as aerodynamic drag, underinflated tires, etc.).

- Now, consider that as your vehicle moves it must **push aside the air through which it travels**. The air *resists* being pushed aside; this is called "**drag**," **which extra ft/lbs of energy are required to overcome**. The *amount of drag* is affected by a number of factors, including the *shape* of your vehicle, it's overall *frontal area* (how bulky it looks from the front), and your *vehicle weight*.

  As we'll discuss later, in addition to the drag of the car itself moving through the air, additional drag is created by all the "stuff" that protrudes from the vehicle, such as side mirrors, roof racks and gear mounted on them, etc. (See "Gas Mileage Techniques For The Road" on page 12).

- Another form of drag is "**rolling resistance**", the **drag created by your tires attempting to roll smoothly over the roadway surface**. Resistance to rolling is a combination of factors, including the *shape* of the tire (fully inflated tires are round and roll relatively easily, whereas underinflated tires are slightly out-of-round and require more energy to roll). *Friction* between the tire tread and the roadway surface (smooth to very rough) is also a factor, as is how well wheels are *greased* where they mount on the axle.

- Finally, friction and resistances within the *engine components* (cylinders and cylinder walls, pistons, and other moving engine parts) also use energy and reduce your mileage for each gallon burned.

It's a good idea to keep these principles in mind when you begin trying to improve your gas mileage.

Now, let's get to the good stuff and look at ways to save money by getting more miles out of every gallon of gas you burn.

First, understand that getting the best gas mileage—and saving money—is *only* possible when you are **CONSTANTLY HIGHLY AWARE** of your driving environment and are very **PROACTIVE** in trying to save gas. You'll seldom get *really good fuel economy* by driving in a leisurely and/or distracted manner. In other words, if you *really* want to improve your gas mileage you must **work at it**. Following are the tools you need to do it, but **only you can apply them to your unique situation**.

Many of the following tips involve **driving more slowly** than you normally do; it will take you longer to get where you're going. So, **plan to leave early** and **allot more time to complete your driving trips**.

Note: New cars won't reach their *optimal fuel economy* until the engine has fully broken in. This generally takes 3 to 5 thousand miles, after which miles per gallon should begin to creep upward.

TAKE CARE OF YOUR TIRES

Riding on **properly inflated tires** is one of the most important keys to achieving good gas mileage. Running on under-inflated tires is a real gas mileage killer, **lowering gas mileage by as much a 0.5% for every 1 psi drop in pressure of all four tires!** Globally, roughly TWO MILLION gallons of gas is wasted EACH DAY due to people driving on under-inflated tires.

Figure A: Recommended tire pressures are shown on Data Placard

Buy a **tire pressure gauge** and **check tire pressures frequently.** Always add air *only* when your tires are **COLD** (early in the morning is a good time). For the best fuel economy inflate tires to **3-5 psi HIGHER** than the recommended pressure (as noted on the **Data Placard** posted on the driver's side door jamb [Figure A], in the glove box and in your Owner's Manual). However, don't inflate tires to more than the maximum pressure shown on the tire sidewall. **NEVER** run tires at pressures **LOWER** than those recommended.

Note that for many cars the recommended pressure for front tires is about 2 psi higher than for the rear tires (this is to accommodate the greater weight in the front due to the engine and all the other "stuff" crammed under the hood).

- To both improve gas mileage and reduce tire wear, periodically have the **ALIGNMENT** of your wheels checked, especially if you feel the car "pulling" noticeably to the left or right.

- Have the **BALANCE** of your tires/wheels checked frequently, especially if you feel a roughness in the ride in your butt (which may mean a **rear tire** needs to be balanced) or you feel shaking in the steering wheel (may indicate a **front tire** is out of balance).

- If you live in an area where you have a *real winter* (you know, snow and such), make sure to change your tires to models suited for the season. Driving in summertime on softer, deep-treaded tires designed for winter use increases tire-to-roadway friction and reduces gas mileage.

Can you identify this important symbol? Here's a hint: it appears on your car's dashboard display, if you have a 2008 thru 2012 year model.

It's a **safety alert light** that warns that the pressure in one or more of your car's tires has fallen to **25 percent lower than the pressure recommended by the manufacturer.** When you see this alert you should check the pressures in *all* your tires, including the spare, and inflate them, if necessary.

But, don't wait until you see a warning: It's a good idea to check tire pressures regularly, at least once a month. According to the National Highway Traffic Safety Administration, tire problems are direct factors in 660 highway deaths and 33,000 injuries every year, and low tire pressures are a contributing factor in many cases. And, as we've mentioned above, **properly inflated tires are a key factor in improving your fuel economy**, as well.

Since tire pressure is so important to both safety and good gas mileage, rather than pay the 75 cents most gas stations now charge to use their air fillup stations, you may want to buy a **digital tire gauge** and invest in a **small, inexpensive 12-volt air compressor** that will enable *you* to add air to your tires when they need it. Available from a variety of manufacturers, most are compact enough to be easily carried in your trunk and light-weight enough to not adversely affect your gas mileage. Prices run from $45 to $200, depending on the features you get (built-in emergency light, operates off of your car's 12-volt cigarette lighter input, ability to jump start your car when you have a dead battery, etc.) Good models will add 10 psi of tire pressure in about 5-7 minutes, and the power cord and air hose store conveniently inside the hard case or a zippered carrying case. Check the Links section at the end of this book for links to several manufacturers of these handy compressors.

Here's a little warning about driving on old tires. Tires deteriorate with age, even if they have never been used, or have been used very little. It is believed that the process of vulcanization, combined with the pressure exerted by a tire's biased steel belts, puts a tire under tremendous internal stress. In addition, oxidation and other degenerative processes that affect the chemical composition of a tire limit its useful life. Experts recommend that tires over six years old, regardless of how much mileage they have on them, be inspected by a qualified technician, even if the tire shows no obvious damage. For best gas mileage, when buying new tires choose **radials.**

Figure B: *Typical portable air compressors suitable for adding air to a low-pressure tire (Photos courtesy Viair Corporation and Campbell Hausfeld Corporation).*

CHECK YOUR TIRE TREAD DEPTH

Although not directly related to improving gas mileage, driving on nearly bald tires is dangerous. Common signs that the tread on your tires is too thin include hydroplaning more than usual on wet roads or skidding when coming to a stop.

Even if you haven't experienced such obvious reactions, for safety's sake it's important to periodically check the depth of the tread. Here's how to do it using a common Lincoln Head penny coin:

Figure C1: *From the top of Lincoln's head to the edge of a penny is 2/32", the same as the minimum tire tread depth.*

Figure C2: *If you cannot see the top of Lincoln's head then your tread depth is more than 2/32". It's a safe tread depth.*

Figure C3: *If you are able to see the top of Lincoln's head then your tread depth is less than 2/32". It's time to buy new tires!*

- The distance from the top of Lincoln's head to the top edge of a penny is 2/32 inch, which happens to be the tire industry's recommended minimum tread depth for safe travel. Use this fact to check your tires...anytime, anywhere.

- Turn a Lincoln Head penny upside-down and insert it in the groove between two sections of tire tread. Then, sight along the tread from the side. If the *top* of Lincoln's head is sunk down into the tread so far that it's not visible (as illustrated in Figure C2) you have more than the recommended minimum tread depth.

If the tread is so thin that the penny doesn't go very deep into the tread, you can see the top of Lincoln's head (Figure C3); this is a warning that you should replace your tires. Again, driving on thin tread is hazardous, particularly on wet roads or in snow.

Be sure to **check all four tires**, and your **spare**.

Tests conducted recently by The Tire Rack, a commercial tire sales company, indicate the penny test may not provide *enough* of a safety factor — you may be driving on nearly bald tires. Instead, they recommend you conduct the test with a *Washington Head quarter* (see Figures D2 and D3 on page 7).

When The Tire Rack (http://www.tirerack.com) compared the stopping distance for cars equipped with tires that passed the penny test (that is, with 2/32 of an inch of tread depth remaining) compared with tires that had passed the quarter test (4/32 of an inch of tread depth remaining...twice the amount of tread), the cars using the tires tested with a quarter took significantly less time to stop than tires with thinner tread, and stopping distances were much shorter. In addition, at highway speed the deeper tread significantly reduced a tire's tendency to hydroplane on wet roadways.

Figure D1: *The space from the top of Washington's head to the edge of a quarter is 4/32", twice the minimum tire tread depth.*

Figure D2: *If you cannot see the top of Washington's head, then your tread depth is more than 4/32". This is a safe tread depth.*

Figure D3: *If you are able to see the top of Washington's head, your tread depth is less than 4/32". It's time to buy new tires!*

So, to be on the safer side, we suggest using a Washington Head quarter to periodically check the tread depth on all four tires on your car.

Be sure to check the depth at several different places on each tire, and on both the inside and outside edges of the tire.

If there's a noticeable difference in tread depth between the inside and outside edge of a tire (particularly the front tires), you may be experiencing uneven tire wear, which may indicate the front end is out of alignment, front end parts are worn out, or your tires are bad. Have your mechanic give the front end a good checkup.

Another way to determine that it's time to buy new tires is when you can see the Tread Wear Indicator Bars that are built into all tires and mark the minimum allowable tread depth. These bars, which run horizontally across the bottom of the tread at several locations, become visible when the tread has worn down to 2/32" and the worn-down tread becomes flush with the bars. Like the penny tread depth test, the Tread Wear Indicator Bars, when flush, warn that it's time to replace your tires. Of course, you can visually check the Indicator Bars at

any time to determine how much tread you have left before it becomes necessary to buy new tires.

PROPERLY MAINTAIN YOUR CAR

- Check for leaks in your gas tank, and to prevent evaporation from the gas tank ensure that the gas cap fits properly. The Car Care Council (http://www.carcare.org/) estimates that up to 147 million gallons of gas evaporate annually due to poorly fitting or missing gas caps. Generally, after filling up, turn the gas cap until it clicks three times.

- Before backing out, check your driveway for evidence of leaking oil, transmission fluid and gasoline.

- Have your car's engine professionally tuned up every 15,000 to 20,000 miles. An engine tuned in accordance with the manufacturer's specifications can improve mileage as much as 3-4%.

- Check to ensure that the fuel/air ratio is in accord with that recommended by the manufacturer.

- Have spark plugs and leads checked and cleaned, and replace as necessary. Have spark plug gaps checked and adjusted, if required. Have the spark timing checked.

- If you drive a car with a carburetted engine, inspect the air filter regularly and clean or replace, as necessary. Doing so can improve fuel economy from 2-6 percent, even more if the filter is so badly clogged that it affects drivability.

Some research indicates that replacing a clogged air filter on cars with computer-controlled, fuel-injected gasoline engines (engines prevalent on most gasoline-powered cars manufactured from the early 1980s to the present) increases *acceleration* but does *not* improve *gas mileage*.

The explanation: a clogged air filter doesn't affect gas mileage because the onboard computer that controls the air/fuel mixture automatically compensates for the reduced air entering the cylinders, making the cleanliness of the air filter basically a non-issue.

Some folks have questioned the truth of the above statement. Since this issue is still a matter of debate, it's recommended that you check your air filter regularly and clean it when it's dirty.

- Clear/clean the fuel filter, fuel injectors and the fuel intake system.

- Fix really serious maintenance problems, such as a **faulty oxygen sensor**. Most cars produced after 1980 have an oxygen sensor, a critical part of the emissions control system that sends data to the Engine Management Computer to help the engine run as efficiently as possible and to reduce emissions.

  Located in the exhaust pipe, the sensor detects the richness or leanness of the air/fuel mixture entering an engine. An oxygen sensor is needed because the amount of oxygen an engine can pull in depends on many variables, such as the ambient air temperature, engine temperature, the barometric pressure, the load on the engine, etc.

  When an oxygen sensor fails, the onboard computer no longer senses the incoming air/fuel ratio and is reduced to "guessing". Performance is compromised, and the engine uses more fuel than is needed. Repairing a faulty oxygen sensor **improve gas mileage by as much as 40%**.

- Check **brakes**, **wheels** and **shocks** regularly. Dragging brakes *destroy* fuel economy.

- When having your oil changed, use the manufacturer's **lowest viscosity** (thinnest) recommended grade of motor oil, preferably a synthetic, or a regular oil that's had a **friction-reducing additive** added. Oil with additives generally say "**Energy Conserving**" on the API performance symbol on the can (as illustrated to the right). Using the recommended oil can improve gas mileage by 1-2%.

SPEED AND GAS MILEAGE

In November, 1973, the Federal Government enacted the National Maximum Speed Law. Enacted as part of the 1974 Emergency Highway Energy Conservation Act in response to rapidly increasing oil prices during the 1973 oil crisis, the law limited highway speeds to a maximum of 55 miles per hour. The purpose: reduce gasoline consumption nationally by 2.2%.

Many motorists simply ignored the new law. Even though the United States Department of Transportation's Office of Driver Research determined total fuel savings to be only 1% (and "independent studies" found a mere one-half of one percent savings), the science behind the law was sound (one of the reasons you save gas at slower speeds is that when your speed is reduced, aerodynamic drag [wind resistance] affects your car less, and your car's engine therefore consumes less gas).

It's interesting that a similar mandated reduction in speed hasn't been enacted as a response to the current ramping up of petroleum prices.

Although speed limits haven't been officially reduced, individual drivers can still employ the science underlying the 1973 law to improve their gas mileage and save money. **Speed is one of the most important factors affecting gas mileage**, and, in general, **when you drive at a moderate speed your gas mileage improves.**

EPA fuel economy tests show that **driving at a steady rate at a fuel-efficient speed definitely improves fuel economy**. The *optimum* speed varies with the type of vehicle. As shown in the graph to the left, **it is usually in the 35 mph to 55 mph range**.

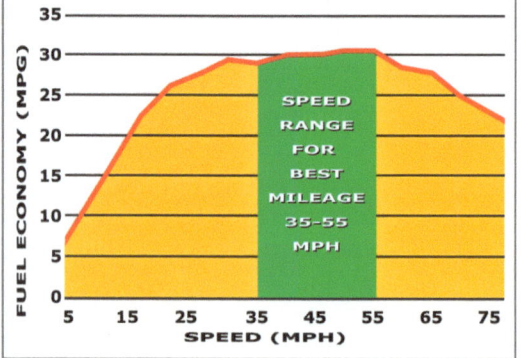

Studies have also shown that speeding and aggressive driving (rapid acceleration and frequent, heavy braking) can reduce gas mileage as much as 33% at highway speeds and 5% in town. If gas costs $3.50 a gallon, your aggressive driving could cost you around $1.16 per gallon.

Note the most fuel-efficient speeds shown on the following chart, which tested a Honda Civic hybrid vehicle with a HCH-II battery pack, traveling a one mile distance at constant speed with Cruise Control engaged and the air conditioning turned **OFF** (Note: The test was conducted on a hybrid, but results should be roughly equivalent for a gasoline-powered car).

Run #	30 mph	**40 mph**	50 mph	60 mph
1	84.3 mpg	**89.3 mpg**	69.1 mph	56.4 mpg
2	86.5 mpg	**89.0 mpg**	64.8 mpg	48.5 mpg
3	85.1 mpg	**90.7 mpg**	67.3 mpg	50.5 mpg
Averages	85.3 mpg	**89.7 mpg**	67.1 mpg	51.8 mpg

As you can see, the **BEST FUEL ECONOMY** is achieved when **cruising steadily in the 35-45 MPH range**, in what I call the **"Sweet Spot"** (See "Find And Take Advantage Of Your 'Sweet Spot'" on page 30 for instructions on how to find and use your car's "Sweet Spot").

Traffic congestion affects speed, and therefore fuel efficiency, as well. Since the best gas mileage is achieved at about 45 mph, fuel efficiency decreases dramatically when traffic is so heavy that cars are forced to slow below 45 mph for extended distances/time. You can increase your mileage by **avoiding roads and times where/when traffic is likely to slow below 45 mph.**

Traveling too fast also reduces gas mileage; for every 5 mph above 55 mph **fuel efficiency is reduced by 7%. You'll get up to 20% better mileage at 55 mph compared to driving at 70 mph.**

In general, for the best fuel economy always travel at the posted speed limit, or slower.

Because of our difficulty, as drivers, in remaining aware of our *actual* speed (and therefore our fuel economy), it's a good idea, when possible, to use your car's **CRUISE CONTROL SYSTEM** (left) **to keep your speed constant**. This not only ensures that your speed remains "fuel-efficient" but also can decrease the number of times you "pump" the accelerator (press down on the accelerator and then let up on it). Repeated "pumping" can significantly lower fuel economy, while a constant, gentle foot on the accelerator greatly improves gas mileage.

> **Using Cruise Control on flat roadways can improve gas mileage by as much as 7%.**
>
> **However, avoid using Cruise Control on hilly terrain.**

One of the greatest problems facing people trying to maximize their fuel economy is that few of us pay constant attention to our *actual* speed. Tests have shown that driving speed is actually a complex issue involving the driver, the vehicle, and road factors. In other words, drivers respond less to posted speed limits and more to a variety of somewhat intangible perceptual cues involving the "feel" of the car, the design of the road, and the physical environment outside their car.

For example, it's much more difficult to keep your speed moderate when a road is straight, wide and flat (you "feel" that you're driving too slowly), compared to one that is curving, narrow and hilly. Regardless of the posted speed limit, most people will naturally drive more slowly on winding, hilly, 2-lane country roads than they will on straight, flat, 4-lane divided highways.

The existence of vertical visual elements, such as trees, buildings and walls, close to the side of the road generally cause people to travel more slowly, since people *perceive* they are traveling *faster* than they would think they were traveling if the roadside was flat and devoid of vertical objects.

People usually drive more *slowly* when car windows are *down* than when the windows are *fully up*. The louder sound and more immediate feel of air moving past the vehicle with the windows down gives the perception that one is traveling faster than the perception of speed one has when the windows are rolled up and the driver hears and feels air passing the car less directly.

Keep these human tendencies in mind as you drive.

GAS MILEAGE TECHNIQUES FOR THE ROAD

- Reduce excessive aerodynamic drag by removing all drag-inducing elements from outside the car: roof racks and carriers, gear boxes, brush guards, running boards, canoes and kayaks, bicycles and bike racks, spoilers (where designed for downforce not enhanced flow separation), large/wide tires, etc. Even though a sleek kayak or a pair of expensive racing bikes look really cool atop your car, the reality is that drag-inducing exterior gear can reduce your gas mileage by up to 5%.

Put gear from the roof in the trunk, instead. If you *must* mount a carrier on the roof, buy one that is **aerodynamically** shaped. Don't leave a roof rack mounted permanently; remove it when your trip is done. If you drive a pickup, *don't* lower the rear lift gate, but *do* install a bed cover. Sun roofs and rough-textured vinyl tops also add unwanted drag.

- A *lighter* vehicle takes less energy to move. So, **reduce your car's overall weight by removing all heavy and non-essential items from the car's interior and trunk** (golf clubs, heavy tool boxes, scuba tanks, junk you've simply been carting around, etc. And, after the last snow, put the heavy bags of road salt in the garage).

  If you have road service with AAA or others, consider taking your **spare tire** and **jack** out of the trunk to lighten the load, and carry a can or two of "flat-fix" to patch leaks from the inside of a tire. You might want to then carry along a DC-powered portable air compressor (see page 4).

- Studies have shown that reducing the total weight of your car by 100 pounds can boost your mileage by up to 2%. This reduction is based on the percentage of extra weight relative to the vehicle's weight and therefore affects smaller vehicles more than larger ones.

- Be aware that towing a boat, trailer or another car naturally reduces your fuel economy.

- With so many of today's drivers being aggressive, just "going with the flow" on the highway can mean driving **10-15 miles per hour over the speed limit**, **which will** *nearly always* **be** *faster* **than the speed that's optimum for your car**.

  Although it's not easy to get accustomed to, for best highway mileage accelerate to the speed at which you get the best mileage (which you've previously determined), then pull into the right-hand lane, maintain that steady-state speed by using Cruise Control, and let the speedsters pass you by (see the section "Find and Take Advantage of Your Transmission's "Sweet Spot"' on page 30). **Using this technique can improve fuel usage by 20% or more**.

- If your car has an **"overdrive" button** (an example is shown to the left), engage it when driving on the highway. Check your Owner's Manual for info on how to use overdrive and at what speeds to engage it.

- On the highway, **maintain your momentum — don't weave in and out of traffic**, and when approaching a hill **add a little gas** to build up speed **before** you reach the hill, then **maintain that RPM** until you reach the crest.

IMPROVING GAS MILEAGE WHILE DRIVING IN TOWN

As we've seen, the speed range at which most cars operate most efficiently (burn the least gas) is between 35-45 miles per hour (see "Speed and Gas Mileage" on page 9-10). Your car's engine is a great deal less efficient at speeds lower than this range. So, when accelerating, it makes sense to get your car up to the most efficient speed as quickly as possible. The technique for fuel-efficient acceleration, then, (particularly from a dead stop) is to add gas at a **MODERATE** rate so that your car accelerates at a steady, but not neck-breaking, rate until you get to around 40 miles per hour (or your transmission's "Sweet Spot").

Never **accelerate quickly or "floor it." Also, don't brake heavily.**

Avoiding "jack rabbit" starts and stops can increase fuel efficiency as much as 33% in town driving.

TRY TO KEEP YOUR SPEED CONSTANT when traveling between traffic lights, anticipate slow downs or stops; never "pump" the accelerator (when your car speeds up and then slows down).

TRY TO USE YOUR BRAKES AS SELDOM AS POSSIBLE. Most drivers use their brakes a great deal more often than necessary. In general, that's because they usually don't leave enough room between their car and the car ahead so that they can simply take their foot off the accelerator and let the speed bleed down naturally (which saves gas). If you drive too close to the car ahead, every time he/she brakes, you must brake, as well.

Here's the problem with braking. *You apply gas* to get your car up to the desired speed, so *the bulk of your gas is consumed in accelerating*. Due to inertia, a great deal *less gas* is used to *maintain* that speed. When you apply the brakes, however, you're burning off the inertia energy you achieved during acceleration into heat energy (due to friction between the brake pads and the brake rotor), essentially *wasting* the energy that burning gas (gas you paid for when you filled up) provided during acceleration.

So, *it's much more fuel efficient to avoid braking*. Instead, **try to leave about 2-3 seconds of *buffer space* ahead (*about 4-5 car lengths*)** so you have the room to "coast" to slow down when the car ahead brakes; this may allow you to maintain a steady speed as much as possible. But, the reality is that this can be difficult to do in town, because when traffic is heavy folks will usually fill the space ahead and you'll be forced to brake, anyway. Just do the best you can.

TRY TO STOP AT TRAFFIC LIGHTS AS SELDOM AS POSSIBLE:

Think about this: **you get ZERO mpg when you are sitting at a red light or are not moving in heavy traffic**. So you really *must* try to **minimize the time you spend idling**. And, **you burn a lot less fuel getting a *rolling* car *moving faster* than is used to get a car moving from a *dead stop***.

SO, THE IDEA IS TO AVOID STOPPING AT TRAFFIC LIGHTS AT ALL.

- As illustrated in the photo to the right, always **LOOK WELL AHEAD DOWN THE ROAD**, and remain aware of the status of traffic lights at least two lights ahead.

- **ANTICIPATE** having to slow down and/or stop for lights that are already red, or that have been green for awhile and are likely to turn red soon (Figures E and F).

In either case, when approaching these lights **take your foot off the accelerator, tap the brake to cancel Cruise Control, let your car *coast*, and try to time your approach so the light turns green *before* you come to a complete stop**.

Always try to keep your car ROLLING (even if it's only moving at 2-3 mph) — it will definitely save gas.

But, this may mean slowing down quite a distance before you reach a light, so be prepared to be honked at frequently since most drivers aren't accustomed to slowing down well before reaching a light, and those behind you may become impatient with your slowness.

Traffic lights are generally timed so that cars driving the speed limit are less likely to have to stop at red lights. A good way to estimate how long you must wait before a light ahead turns green is to check the amber "countdown timer" mounted on a corner of an intersection; these will show an amber "hand" icon (Figure E, below), indicating that you still have 30 seconds or more before the light changes, or will be counting down in seconds (Figure F) until the change — in this case the red light will change to green in 11 seconds. **LOOK AHEAD, AND USE THESE INDICATORS TO AVOID STOPPING!**

Figure E: *"Don't Walk" signal at a traffic light indicates that the red light will not change to green for 30 seconds or more.*

Figure F: *"Don't Walk" signal with a "countdown timer" at a traffic light indicates the red light will change to green in 11 seconds.*

Here's an easy and effective trick to avoid consuming the extra gas you'd normally burn to get your car rolling again after stopping at a traffic light or stop sign:

Most folks know that if they are driving a car with an automatic transmission and they take their foot off the accelerator while in gear, the idle RPM of the engine will often cause the car to slowly start rolling on its own without the driver applying any gas at all.

So, when you are ready to get moving again after a full stop simply take your foot off the brake pedal and let the car accelerate on its own for a few seconds before you apply gas.

Remember that it takes a lot more energy to get an object at rest moving than it takes to get a moving object moving faster. Using this idea, make use of your engine's ***idle RPM*** to gain some movement, so you're not using any *extra* gas to gain this small amount of initial speed (the engine is running at idle RPM anyway). This is the same concept as not allowing the car to come to a complete stop when approaching a traffic light. Just keep the car rolling.

COASTING SAVES FUEL…BUT

Here's a technique you can borrow from avid hypermilers which, when applied intelligently, can help increase gas mileage. It's called **COASTING**.

When approaching a red light, or one that may soon turn red, rather than stepping on the brakes consider reducing speed by slipping your transmission from **DRIVE** to **NEUTRAL** (or simply depressing the clutch pedal if you have a manual transmission) and allowing your car to **COAST.** This technique discontinues the application of kinetic energy from the engine (obtained by burning gasoline or diesel fuel) through the transmission to the differential, and allows aerodynamic and rolling drag (which, when the car is in gear, is overcome by energy supplied by the engine) to slow down your now unpowered vehicle. This will save gas. Here's why.

Even if you take your foot off the accelerator, fuel continues to be supplied to the engine, just less of it. As a result, the engine RPM will drop to IDLE RPM, generally around 900-1000 RPM. This reduces fuel consumption. Then, if you slip the transmission into NEUTRAL, the RPM will drop a bit more, often another 200 RPM or so (saving even more gas). In cars with fuel-injected engines, the Engine Control Unit (ECU) will cut off the fuel supply, but the engine will continue turning over due to the wheels turning. Then, you can use judicious braking to maintain some rolling motion until the cars stopped at the red light ahead begin to move when the light turns green.

Be aware that coasting in neutral is illegal in most states, so if you choose to coast, use the technique judiciously.

USING YOUR AIR CONDITIONING SYSTEM

Most of us are aware that using our car's air-conditioning system reduces gas mileage. Here's why.

When the air conditioning is turned ON, about 5 horsepower of the energy generated by the engine is used to operate it, and the engine must then supply additional horsepower to maintain any given speed (this additional horsepower is not required when the air conditioner cycles OFF, which it does periodically). The result: more fuel is consumed when the A/C is turned ON.

So, use the air-conditioning only when absolutely necessary, since doing so can reduce gas mileage by **10-30%**.

Here are a few tips for getting the best mileage when using your car's air conditioner:

- If you live in an area where summers can be brutally hot, when buying a new car choose one with **light-colored paint**. The light color will reflect the sunlight, the interior will heat up less, and you will be comfortable running the air conditioning at a lower setting.

- To keep the interior of your car from heating up, **park in the shade** as often as you can and use a **window shade** on at least the front windshield. The air conditioner won't have to work as hard to get the inside temperature down to a comfortable level, and the window shade protects the dashboard, interior fabrics and plastics from deterioration due to ultraviolet rays.

  It's also a good idea to get **window film** on all windows (in many states it's illegal to put window film on the windshield itself; use a window shade instead). To really keep your car's interior cool, ask for film that contains **micro-thin layers of metal**, which act as infrared (heat) reflectors/absorbers and visible light filters. These layers are so thin that they are essentially transparent.

- Parking in the shade reduces "**evaporative emissions**": when fuel in the gas tank heats up, it expands and vaporizes, and contains both liquid gasoline and gas fumes; rising pressure in the tank causes gasoline vapor to be forced out of the gas cap and into the atmosphere. So, you're not only losing gas, but you're polluting, too. Tighten your gas cap **at least three clicks** to ensure it seats properly.

- When you get into your car on a hot day roll all the windows **DOWN** to let out accumulated hot air. Once you're moving, **keep the rear windows partially down** for the first 5 minutes or so to pull out any remaining hot air. Then, **if you will be traveling at 50 MPH or more, roll ALL the windows UP and turn the AC ON**.

- When it's hot outside, the alternative to turning on the A/C is to simply roll down the windows. But doing so compromises the aerodynamic sleekness of the sides of the car (intentionally designed to reduce drag and conserve gas) and allows air under pressure to enter into the inside of the car. This increased drag has the effect of slowing the car. To maintain a given speed, more gas must be supplied to the engine, reducing fuel economy.

But, when you are not driving very fast, aerodynamic drag is a less significant factor in fuel consumption. So, when driving slowly around town on a hot day, to save gas **ROLL DOWN all windows**, **turn the A/C OFF**, and **adjust your car's ventilation controls to allow outside air into the cabin**.

- On the other hand, when you're driving at **highway speeds (50+ mph)**, aerodynamic drag is much lower when the windows are rolled UP, and gas mileage improves. When the windows are DOWN at high speed, air pours into the car's cabin and creates a lot of aerodynamic drag. The additional energy the engine must then generate to overcome drag and maintain speed can be **SIGNIFICANTLY MORE** than that required to operate the air conditioning.

  To feel for yourself how aerodynamic drag affects a car at highway speed, the next time you're on the highway roll the window down and stick your hand out into the slipstream (you've probably done this countless times as a kid). All the wind resistance you feel against your hand (and the car) must be overcome by your car's engine, which it does by adding more gas.

  So, at highway speed you'll save gas by **turning the A/C ON**, **rolling the windows UP tight**, and **adjusting your ventilation controls** to **RECIRCULATE** the cold air generated by the A/C within the cabin.

- When the outside air temperature is not *too* hot, rather than turning on the A/C adjust your **ventilation system** to let cooler outside air flow into the car. It will help to roll down your rear windows an inch or two to help pull warmer air out of the cabin and reduce the backpressure on the cooler air that's trying to enter the cabin through the ventilation system.

  Most air conditioners have an "economy" setting that allows for the circulation of unchilled air. Many also have a "maximum" or "recirculation" setting that reduces the amount of hot outside air that must be chilled. Both settings can reduce the air conditioning load — and save gas. Check your Owner's Manual on how to use these features.

- When stopped at a stoplight, railroad crossing, etc., **turn the A/C OFF to** *reduce* **engine RPM about 200 revs per minute**.

  To recap:

- When weather conditions permit, keep the **A/C OFF** and the **windows rolled UP**.

- **When traveling slowly**, roll the **windows DOWN** and **turn the A/C OFF**.

- **When traveling fast**, reduce drag by **rolling the windows UP** and **turning the A/C ON**.

USE THE GRADE OF GAS RECOMMENDED BY YOUR CAR'S MANUFACTURER

Your car's manufacturer recommends the most effective octane level for your engine. For most cars, this is **Regular unleaded gas** with an **Octane Rating of 87** (High Test has a **93** Octane Rating).

To find out the recommended octane for your car, check in your Owner's Manual; it will be listed in the Index under "Gasoline".

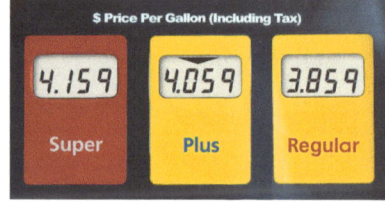

Roughly one in five drivers buy High Test, thinking that they'll get better mileage. But High Test costs roughly 20 cents more than Regular, and less than 5 percent of cars have the high-compression engines that *require* high octane fuel (usually highly-stressed engines in Ferraris, Lamborghinis and Maseratis). In general, there's no benefit in using a higher octane gas than the manufacturer recommends: it won't make your car perform better, go faster, get better mileage or run cleaner.

If your car *does* start knocking, switching temporarily to the next highest octane — mid-grade or premium-grade — gasoline will likely eliminate the problem. If the knocking or pinging continues after one or two fill-ups, you may need a tune-up or some other repair. After the tune-up, you should be able to go back to using the octane grade your car's manufacturer recommends.

Unless your engine is knocking or pinging on Regular, arbitrarily using higher octane gas is a waste of money.

> **Note: Only use E85 Ethanol fuel (85% anhydrous ethanol, 15% unleaded gasoline) in cars designed specifically for that fuel, *never* in a car designed to burn regular gasoline. In fact, the use of E10 "gasohol" (10% ethanol and 90% gasoline, currently the most comm-**

only available retail automobile fuel) in cars manufactured before 1980 can, over time, result in faster-than-normal deterioration of many components of the automobile. Spark plugs, catalytic converts, oxygen sensors, etc. may all deteriorate more quickly. Rubber parts such as gaskets and seals may be weakened or destroyed over time. Engine oil should be changed at shorter intervals. Much of the damage to components results from the higher concentration of water in gasohol compared to gasoline (gasoline *repels* water, ethanol *readily absorbs* water), which tends to literally rust gas tanks and other metal components.

SEARCH OUT WHERE TO BUY THE LEAST EXPENSIVE GAS IN YOUR AREA

Obviously, the less you pay for gas in the first place the less your monthly fuel bill will be. And, since almost all of today's gasolines contain essentially the same ingredients, there's no reason to worry about even the least expensive gas harming your engine.

In fact, although major brands of gas (Shell, Texaco, Chevron, etc.) contain detergents that keep valves, fuel injectors and some other critical parts of an engine working properly, the Environmental Protection Agency *mandates that even the lowest-priced gas contains a specified minimum of detergents*, as well. If you want to bump up the amount of detergent in your cheap-fuel tank, buy a can at a discount auto parts store and add it yourself for a little over a buck. Or, simply buy a major brand gas occasionally to take advantage of its greater detergent load.

Either way, you'll be saving money compared to buying High Test gas.

Developing a strategy to consistently find the least expensive gas is *critical* to keeping your fuel costs reasonable. Here are some suggestions:

- Arm yourself daily with information about current gas price trends by **actively** and **consistently paying attention to the price of gas at stations** you drive past during the day. Carry a small **tape or digital recorder** so you can keep a running record of prices at various stations. When you get home, record this info on your computer to get a good idea of what current prices are, where prices are trending, and where the cheapest gas can be found along routes you travel frequently.

- You may pay less for gas if you **fill up at least three days prior to a holiday**, before gas prices are traditionally increased.

- If gas prices are increasing rapidly in your area — going up a few cents every few days — overcome the psychological tendency to not buy because a tankful costs so much. Instead, **buy often. Don't wait until your tank is nearly empty to fill up**. Instead, buy gas **when you are down to three-quarters of a tank**. That way, you are buying when gas prices have only gone up a little.

    On the other hand, if **gas prices are dropping**, burn through about **3/4 of a tank** and then fill up at the lower price. Besides, driving on a nearly-empty tank can improve your gas mileage, since your car is lighter (you're not hauling around the added weight of a full load of fuel).

    Never let your fuel level get so low that **you're forced to buy whatever gas is immediately available so you won't run completely dry**. Think about gassing up when your gas gauge shows that you have about a **quarter of a tank remaining**.

- Gas stations generally change their prices sometime during the day, so if prices are consistently going up, as they tend to be these days, if you buy gas **early in the morning** you still may benefit from yesterday's lower prices, and, before prices are jacked up later in the day.

    Have you ever wondered why gas prices at a station you pass on the way to work in the morning are *several cents higher* when you head home in the evening. After all, it's the *same gas that was in the station's tanks in the morning*! The reason: station owners often hike gas prices based on information they receive from refineries and distributors concerning ***pending price increases***. When an owner is told the price per gallon of the *next* delivery to his station will increase by a specific amount, he or she may immediately increase the price of the gas that is already in the tanks so they'll have the money to pay the increased cost of the next delivery.

- When traveling major highways or the Interstate, you may be tempted to buy gas at stations along those routes — you know, at the **super-stations** with 20 pumps and a huge, affiliated stop-and-go grocery on site. **Don't do it!** Prices at stations along Interstates and major highways, and in the vicinity of vacation destinations, are often **significantly higher** than at smaller stations near residential areas.

  If you're on a highway trip, consider filling up when you reach a moderate-sized city, where you can take advantage of off-the-major-highway locations and competition between a number of gas stations to find lower fuel prices. Pull off the freeway and head toward the central business district; along the way you'll likely find prices dropping.

  Likewise, if you've traveled a long, empty stretch of highway with no gas stops, the first station you finally come across will likely charge more for gas than those a little further down the road (you know, many folks may be getting a little low on petrol and will pull in at the first gas available... and station owners know it). So, go a bit further and you'll likely pay less. Even better, don't patronize stations that are right on Interstates or major highways; instead, drive off the major highways a couple of miles and you'll find that gas prices drop even further (don't drive too far afield and waste gas, though; if you have a smart phone, use your gas apps to check prices/locations/directions in the area in which you're driving — see page 25).

  And, on a road trip be an opportunist: fill up whenever/wherever you stumble across gas that's significantly cheaper than at other stations you've passed, even if you don't need gas at the moment. Don't wait until you're down to a quarter of a tank since the cheap gas may not be available then.

- Live fairly close to a **Costco** or **Sam's Club** store? Consider buying an **annual membership** so you can take advantage of the lower prices at these big-box stores, where per-gallon prices for Regular gas are often 5-7 cents lower than at other area stations. But, be sure to **check their prices online for a few weeks before buying a membership**, since discount store prices are not always *significantly* lower. If their prices are consistently lower during the test period, and you don't have to drive too far from your normal route to get there, paying the annual membership fee might be a wise thing to do.

- Some gas stations offer a discount to customers who **pay with cash**. Check the pump for a cash price, or ask the clerk.

- Consider using your **credit card** to purchase gas if your credit card provider offers a "dollars back," rebate or "points" program on purchases. However, be sure to carefully read the fine print governing these programs, since some credit card companies limit the reward or rebate amounts customers can earn. But, you'll enjoy a pretty good deal if you are able to purchase gas at a relatively low price *and* earn reward points at the same time. *Sweet*!

- **Gas Rebate Cards** are available that reduce gas purchases by 20-40 cents a gallon. The introductory rebate on many cards is fairly substantial — 10% or so — with long-term rebates dropping to 3-5%. Even that's not bad: If your car gets 20-mpg and you drive 15,000 miles a year on $4 gas, a 5% rebate could save you about $150.

  Each card has a different offer, and different requirements. Some have annual fees, and some require that you buy gas from specific vendors – such as BP, ExxonMobile or Shell. When choosing a gas rebate card, treat it as you would any credit card, and read — and be wary of — the fine print.

  Using one of these cards, it's important that you pay off the amount owed monthly; otherwise, interest charges could wipe out the amount the rebates save you.

  Rebate on some cards expire after a time, and you should find out how rebates are paid by the card(s) you're using: some cards send monthly or annual checks; some apply your rebate to future purchases; and others let you accumulate points to be redeemed for cash, airline miles or prizes.

  For more information of gas rebate cards, and for help sorting out the details of a number of gas rebate card offerings, check out the following websites:

  http://www.creditcards.com/gas-cards.php

  http://www.cardratings.com/gasrebatecreditcards.html.

- Check at gas stations along your normal routes to find out if they offer **discounts on certain days of the week**. If they do, make a note to top up on those days.

- Check at stations for **gasoline discount membership clubs or cards** and consider signing up.

Use the Internet and your SmartPhone to help locate the cheapest gas near where you are:

- The **Internet** and **dedicated Smartphone apps** make it easy to find the lowest gas prices in your local area. Here are some full-featured **websites** that track local gas prices daily on the Internet:

  - www.gaspricewatch.com — Lets you create a "route" you drive frequently and displays gas prices along that route.

  - www.gasbuddy.com — One of the most popular gas price monitoring sites.

  - www.fuelly.com — Lets you compare fuel economy with other drivers, and makes and models of cars, and its mobile phone interface lets you compute your gas mileage right at the pump.

  - American Automobile Association (AAA) Tripkits keep track of changes in local gas prices.

If you're on the road, you can pinpoint the cheapest gas in the neighborhood using an **iPhone** or **Android SmartPhone** and an **appropriate app** that uses GPS technology. These apps first determine your current location, then search a database of all stations near you; they then display the station name, current price and a map to help you find the station. Plus, if you happen across a station enroute that has even lower prices, you may be  able to add it to the database and alert your friends. There are a number of apps out there; some charge a modest price but many are free:

- GasBuddy — this free app for iPhone, Android and Windows 7 offers the choice of displaying stations as a list or on a map, and by proximity to your location or by price. You can also sort by grade of gas (Regular, High Test, Diesel). One of the most full-featured gas apps. User reviews are generally very positive. On the web: **http://gasbuddy.com/**

- iGasUp — (Price: $0.99 for first year, $2.99 per year thereafter) — uses the iPhone 4S's GPS information and the Oil Price Information Service (OPIS) database — used by MapQuest, Garmin, AAA, Sirius Satellite and other technology leaders — to locate the 10 cheapest gas stations near your location with prices displayed low to high and accompanied by the address and brand of the location. You can even look up the cheapest prices within a specific zip code. Users of the application can have their search criteria include regular unleaded gasoline, premium unleaded or diesel, and soon will be able to view station amenity information (Car Wash, Convenience Store, Service Center, etc.). Some customer reviews have been pretty negative. Available via the Apple App store. On the web: **http://www.igasup.com/**

- **CheapGas** — once you've found the gas station you want to visit, this free app for iPhone, iPod, iTouch, and iPad provides a mapped route to get there. Stations are sorted by price. Station pricing is provided by GasBuddy.com (so why not use the GasBuddy app, instead?). Customer reviews of version 3.03 have been negative. On the web: **http://itunes.apple.com/us/app/cheap-gas/id290765007?mt=8**

- **POYNT** — free app that runs on iPhone, Android, Blackberry, Windows Phone 7 and Nokia. POYNT displays local gas stations near your current location, listed by price from lowest to highest. Unlike many gas apps, station prices are updated DAILY. You can even call a selected station. In addition to gas pricing and locations, POYNT also lets you locate your friends, local businesses, retailers, restaurants and events, see what movies are playing by title, theatre, or genre and view movie trailers and reviews, get the weather, and get directions from where you are to anywhere. Rated as one of the top ten apps for the iPhone. On the web at: **http://poynt.com/**

- **GasHog Fuel Economy Tracker** — (Price: $0.99). Unlike the other apps described here, GasHog (runs on the iPod and iPod Touch) does *not* find the stations with the cheapest gas for you; rather, it lets you input the gallons of gas you pump and the miles driven since your last fill-up and the app then calculates the fuel economy of your last tankful. It can also display historical averages. GasHog also offers tips for improving your gas mileage. GasHog does *not* require Internet connectivity to use any of its features, and operates in areas with no mobile or WiFi coverage. On the web: **http://itunes.apple.com/us/app/gashog/id284957432?mt=8&i=6372516**

Drive a particular route regularly? Keep track of prices at various stations by going online *before* you leave home, or check prices using your SmartPhone, then *verify* prices as you pass or stop in at a station. Do this for a few weeks and you'll start noticing pricing trends, and will likely be able to predict when prices will increase at particular stations. All valuable skills that can save you money!

Unless you know a particular route you'll be driving and you've checked the prices at gas stations along that route, don't be tempted to drive too far to buy cheaper gas — you'll just burn extra fuel getting to the station that's selling the cheaper stuff. In general, **locate a source of cheap gas near your home, where you work or where you shop**.

KEEP CONSISTENT RECORDS OF YOUR FUEL ECONOMY

You won't be able to determine whether your gas mileage is improving or not unless you keep track of how many gallons of gas your car is **actually using** to go a certain distance. That requires that you put in a little extra work at the pump — rather than simply filling up, chatting with the blonde who's filling up the Jaguar convertible at the neighboring pump, and then cutting out.

Since I'm *always* interested in what gas mileage I'm getting with my car (not just when there's a spike in the price), I'm interested in how my car's mileage compares with the mileage other folks are getting. So, at the filling station I make it a point to notice whether other customers are writing down the details of their fillup in a notebook. I find that, in general, very few are. And, I'll often ask other folks what mileage they're getting in their make and model of car, especially if they're driving one of the cars I see being advertised as being particularly fuel efficient. Usually, they admit they don't know what their mileage is.

How to calculate your actual miles per gallon (it's actually a lot easier than it sounds):

Always **write down** the following specifics when you fill up (keep a notebook in your glove box):

- The **date** (you may also want to write down the **location** and the **price per gallon**).

- **How many gallons it takes to fill up completely**: read this figure directly off the pump.

- **How many miles you've driven since the last fillup**. Read this from your car's **odometer.**

In order for this system to work, you must do two things:

- **KEEP TRACK** of the miles traveled between fillups: **reset the odometer** on your instrument panel to **zero** before you leave the pump.

If your car doesn't have an odometer, write down the mileage shown on the **total mileage** indicator on your car's instrument panel when you fill up, and then **subtract** that mileage from the miles shown when you *next* fill up. The difference is the number of miles you've traveled between fillups.

- **Fill up COMPLETELY each time you refuel (until the pump shuts itself OFF)**. This is the *only* way you can determine the *actual* number of gallons of gas you've burned between fillups. To ensure consistency in this part of your calculation, always add gas just until **the pump shuts OFF automatically**, then don't add any more fuel (please don't overfill ["TOP OFF"] and spill gas).

Note: After the pump shuts OFF, turn your hose nozzle upside down (with the nozzle still in your fuel filler pipe); **this allows about 1/2 cup of additional gas to drain from the hose** — this is gas you've paid for that otherwise wouldn't find its way into your tank!

Don't forget to reset the odometer to 000 after filling up. If you forget, forego calculating your mileage until after the *next* fillup.

Once you have these two pieces of information you can calculate your gas mileage for the distance you've driven between your last fillup and this one. Here's how:

- **Divide** the total number of miles traveled between fillups (from the odometer) by the number of gallons required to fill up (from the pump); this is your fuel economy in MPG (miles per gallon). Keep a log of your mileage in a notebook in your glove box. Here's an example:

- The **odometer reading** when you pulled into the gas station reads: **296 miles**

- The **number of gallons** it took to fill up completely (as shown on the gas pump): **14.6 gallons**

- **Divide** 296 miles by 14.6 gallons:

> 296 miles / 14.6 gallons = 20.3 miles per gallon

So, you've traveled an average of **20.3 miles** on every gallon of gas you've burned between these fillups. Not too bad if you've got an older eight-cylinder model car and most of your driving has been short trips within the city. Your mileage should be much better if most of your driving has been at 45 miles per hour out on the open highway.

And you can see how easy it is to calculate your mileage once you're armed with the gallons you've use and the miles you've driven. Piece of cake! Now you are part of a small percentage of drivers who actually *know* how fuel efficient their car really is!

It's a good idea to ask folks who drive cars like yours what mileage they're getting. If their mileage is roughly the same as yours, you can be pretty sure that you are checking your mileage properly, and that your car is properly tuned up. Don't expect complete agreement, however, because many variables between how you and others drive can result in differences in the mileage you realize.

More importantly, if your mileage decreases significantly over time, or is a great deal less than the mileage others driving cars similar to yours are getting, you should have your favorite mechanic check for a possible problem with your car.

WHAT ABOUT COMMERCIAL GAS-SAVING DEVICES?

Be very skeptical about any gizmo that promises to improve your gas mileage. The U.S. Environmental Protection Agency has tested devices that are advertised as being capable of significantly reducing your fuel consumption — including "mixture enhancers" and fuel line magnets — and found that very few provided any fuel economy benefits. Those devices that did work provided only a slight improvement in gas mileage. In fact, some of these products may even damage your car's engine or cause a substantial increase in exhaust emissions.

One category of devices that have gotten good reviews from users I've talked to are COLD AIR INTAKE SYSTEMS, which direct cooler, denser air into the engine's air intake, supposedly increasing gas mileage.

For a full list of tested products and their results, visit **www.epa.gov/otaq/consumer/reports.htm**.

FIND AND TAKE ADVANTAGE OF YOUR TRANSMISSION'S "SWEET SPOT"

There's a **SPECIFIC SPEED** at which every car's transmission **SHIFTS INTO HIGH GEAR OR OVERDRIVE** (on 5-speed manual and 4-speed automatic transmissions); at this speed the **RPM** (Revolutions Per Minute) **DROPS DRAMATICALLY** and the engine uses **LESS GAS** while maintaining that **SAME SPEED**. Basically, at this speed/RPM combination **LESS GAS** is used to power the car along at a moderate speed.

You'll noticeably improve your fuel economy if you take advantage of this "**SWEET SPOT**." Here's how to figure out what the "Sweet Spot" is for your vehicle:

- **NOTE:** It's easier to determine the "Sweet Spot" if your car has a **TACHOMETER**.

    As your car accelerates, carefully watch the speedometer as you approach 40 MPH, since most cars will shift into high gear **BETWEEN 40-45 MILES PER HOUR**. When **THE TRANSMISSION SHIFTS INTO HIGH GEAR** you'll see **THE TACHOMETER READING DROP BY 800-1000 RPM**.

FIGURE G: *The **tachometer** (left) and **speedometer** (right) showing their respective readings when my Jeep Cherokee is **about to shift into high gear**. The **speed is 40 MPH** and the **engine is turning at 2000 revolutions per minute**,* as indicated on the tachometer.

- Figure G shows the tachometer and speedometer on my automatic transmission 2000 Jeep Cherokee when the car reaches 40 MPH. Note that the tachometer shows that the engine is turning at 2000 RPM to maintain this speed.

FIGURE H: *Readings on my Jeep Cherokee's tachometer (left) and speedometer (right) as the vehicle accelerates a bit, just before the transmission* **shifts into high gear and reaches the "Sweet Spot."** *The speed is* **45 MPH** *and the engine is turning at* **2200 RPM.**

- Figure H shows the tach and speedometer after I accelerate from 40 to just under 45 MPH, the speed at which the automatic transmission in my Cherokee shifts into **HIGH GEAR**. Note that the tach reads 2200 RPM, up 200 RPM from the 2000 RPM it indicated at 40 MPH (Figure G).

FIGURE J: *In my Jeep Cherokee, when the RPM reaches 2200 RPM and the speed reaches 45 MPH, the* **engine RPM DROPS dramatically**, *as shown on the tachometer (left),* **from 2200 RPM to 1350 RPM (the "Sweet Spot")**, *while the speed on the speedometer (right)* **remains at 45 MPH**. *If I now* **engage Cruise Control and automatically maintain speed**, *my Jeep will be traveling at a moderate speed at an RPM (and fuel consumption level) that is only* **marginally above the engine's 900 RPM idle rate.**

- As shown in Figure J, the instant that I apply a bit more gas and the engine **shifts into high gear, the RPM will DROP from 2200 to 1350.** In other words, **45 MPH is the speed at which my Jeep's transmission shifts into the "Sweet Spot" and the RPM drops to just a few hundred RPM above idle. At this "Sweet Spot," my Cherokee is traveling at 45 MPH using relatively little fuel.**

- So you can **maintain the "Sweet Spot" RPM at which the engine uses the least amount of gas for the best speed**, be ready to **immediately engage the CRUISE CONTROL the instant the RPM drops.**

- With the engine now running at a **REDUCED RPM**, with all other factors being equal **YOU WILL BE SAVING GAS.**

- The "Sweet Spot" speed and RPM combination for your car will likely be a little different than in my Jeep, but the speed should be somewhere around 45 MPH.

- If your car lacks a tachometer, you can easily determine the "Sweet Spot" speed **simply by noting the speed at which the engine RPM drops noticeably** — again, around 45 MPH. Even though you won't know how much the RPM drops when your transmission shifts into high gear, you can be sure that it will be significantly lower.

  To determine your car's "Sweet Spot" speed without a tachometer, take your car out on a smooth and relatively quiet stretch of road and you should be able to hear and feel when the tranny shifts into high gear.

- Since 40-45 miles per hour is **SLOWER** than most folks like to drive, be prepared to pull into the **right-hand (slow) lane** and let everyone else pass you by. Also be prepared to get **honked at** fairly frequently…simply flash a smile as other cars pass you. You'll likely get to the next light only a few seconds later than the speeders, and you may even pass them when they pull into the next gas station to fill up.

  NOTE: this technique works best when traffic conditions allow you to **travel at a steady speed for some distance** without having to slow down or stop for lights (highway driving). It can be difficult to use this procedure when driving in the city, where it's often "stop-and-go" between stop lights.

Use your car's Fuel Economy Gauge to verify your gas mileage

In addition to finding and utilizing the advantages of your car's "Sweet Spot," if you have a late-model car with an electronic instrument panel you should also pay close attention to the information provided by your car's Fuel Economy Gauge (FEG) (Figure K). The FEG lets you see in real-time how your current driving is impacting your overall gas mileage.

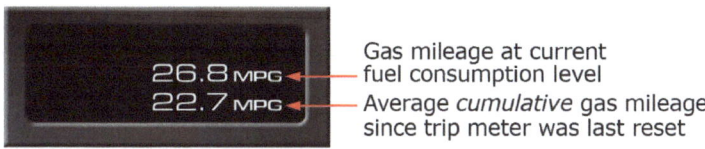

26.8 MPG ← Gas mileage at current fuel consumption level
22.7 MPG ← Average *cumulative* gas mileage since trip meter was last reset

Figure K: *A typical Fuel Economy Gauge in a new car that provides information to the driver concerning both the number of miles per gallon (MPG) realized at the current speed (top) and the cumulative, average mileage realized since the car's electronic odometer (trip meter) was last reset (bottom). This straight-forward Fuel Economy Gauge is used on a 2010 Subaru Outback.*

Use the FEG to cross-check the gas mileage you calculated yourself using gallons burned divided by mile driven (see page 28), and for verifying that the speed at which you believe your car shifts into the "Sweet Spot" is, in fact, the speed/RPM combination that gives you the best fuel economy; read this on the "Gas mileage at current fuel consumption level" readout. In other words, use **all the tools at your disposal, in combination**, to determine **how you must drive your particular vehicle to coax the very best gas mileage from it**. This may mean changing the speeds at which you drive to realize the best fuel economy, or changing the pressure in your tires, or...

If your car doesn't feature an electronic Fuel Economy Gauge, you can still take advantage of having one onboard your car by installing one of a number of **after-market Fuel Economy systems**.

Momentum-based fuel-economy gauges sense how rapidly you're accelerating and braking and display excessive or acceptable fuel consumption levels by means of red and green lights or gauges.

Image of ScanGauge II Fuel Economy Meter courtesy of Linear Logic LLC

More current digital Fuel Economy Gauges read your car's **diagnostic port (OBD II)**, which is standard on all American- and Canadian-built vehicles sold since 1996. The OBD II lets aftermarket Fuel Economy Gauges access data from the vehicle's **Engine Control Module (ECM)** so that a number of functions, including real-time mpg, are displayed to the driver.

Some after-market gauges are designed to work with *all* OBD II-equipped cars, while others work only on vehicles that use a **Controller Area Network (CAN) electrical system**, which was introduced in new cars in 2003 and is now standard on most new vehicles. Before purchasing an after-market Fuel Economy Gauge system for your car, ascertain which system your car is equipped with. Cost for these systems range from around $70 on the low end to over $330 (not including the cost of mounts and mounting accessories).

See the "Links" section on the last page for a partial list of manufacturers of after-market Fuel Economy Gauges.

"PLAN" TO GET BETTER FUEL ECONOMY

When heading out to do "chores" in your car, intentionally PLAN YOUR TRIPS so you get the best mileage possible, and then stick with your plan.

If you're a multi-car family, use the vehicle that gets the best mileage as often as possible. Unless you must haul lots of passengers or "stuff," avoid using your SUV or mini-van. If you have a larger vehicle, spread the cost of using it around the neighborhood: invite a neighbor or two to join you on your trip to the grocery, taking the kids to the Little League game, etc. Ask them to share the cost of the gas you burn.

Plan to get as much done in a SINGLE TRIP as possible.

Intentionally seek out the best route, one that avoids heavy traffic, railroad crossings, schools and traffic lights. Remain aware that **the shortest route is not necessarily the one on which you'll save the most gas**, particularly if you must stop-and-go a lot.

To help with trip planning, you may want to make use of **a moderately-priced Global Positioning System (GPS) device** to map out the **SHORTEST, MOST DIRECT ROUTE** to your destination. On the road, use the GPS to remain on the route. It will help you both save time and avoid burning gas wandering around lost.

If you don't have a GPS device, consider using **www.mapquest.com** to help you determine which destination is the furthest from your home, and to map out the best route to get there. To use MapQuest:

- Enter the address of the destination at which you will **END** the trip, then press the "Get Directions" button.

- Enter the address of the location from which you will **START** the trip. Press the "Get Directions" button again.

- Mapquest then displays the suggested route to your destination as a blue line on the map.

- Route directions in discrete steps, with the time, mileage and estimated fuel cost of each segment, are displayed on the left of the screen.

As you plan your route, count how many stop lights, train tracks, school zones and drive thru's you sit at during a normal day.

- Drive a different route to/from work, for example. Try taking back roads rather than major routes, as these often have less traffic and you may need to stop-and-go less often.

- Avoid driving during rush hour: stop-and-go traffic is a real gas waster, so negotiate with your employer to let you arrive at work early (or late) to avoid the morning traffic crunch, and leave work early (or late) in the afternoon. Pay attention to traffic volume in your areas, and do your shopping and run errands at times when there are fewer people on the road.

- Consider doing your shopping **at night**. That way, you'll likely travel in lighter traffic with less stop-and-go, and will probably spend less time driving around parking lots looking for an empty spot.

Your engine operates most efficiently once it's warmed up; that's one reason your fuel economy drops when most of your trips are short ones — the engine simply hasn't had time to heat up to the ideal operating temperature. Running short trips, each from a cold start, can burn twice as much gas as a single trip covering the same distance when the engine is warm.

So, **plan to do the first of your "chores" at the location that is FURTHEST from your home or office**, then move progressively closer to home; this give's your car time to warm up to and maintain the ideal operating temperature during the shorter segments of your drive around town.

Don't start your car until you are actually ready to leave; letting the engine idle while waiting for others to get their gear together can waste a lot of gas.

NOTE: sitting in the driveway for a prolonged period waiting for the engine to warm up does **NOT** help your gas mileage; in fact, doing so actually consumes *more* gas, wastes time and increases pollution. **Don't warm up your engine for more than a minute.**

Another note: **Before shutting down the engine, turn OFF all electronic equipment** (air conditioning, radio, disk players, GPS, etc.) to lessen the load on the engine when you start the car the next time. Starting with the air conditioning ON, in particular, makes your engine work harder, but the cumulative effect of having to power on a variety of electronic components also burns more fuel.

And, if you **REALLY** want to get serious, consider doing what FedEx and United Parcel Service (UPS) are **REQUIRING** their delivery drivers to do to conserve gas: plan your local trips to the grocery, the pharmacy and the post office **so you only make RIGHT TURNS at intersections**.

The reason: **MINIMIZING IDLING TIME. For every hour you sit at idle you're burning a about a gallon of gas**. During a day on the road vehicles waste a lot of time—and gas—waiting for left-turn signals to come on at traffic lights, or to make left turns across oncoming traffic; on the other hand, since right turns on red after a complete stop are legal in most states, making right turns requires significantly less idling and burns less fuel.

Whether or not you go to the extreme of making right turns only, ***DO CALCULATE HOW MUCH TIME YOU SPEND IDLING EACH WEEK, THEN TRY TO REDUCE IT!***

Every year billions of dollars — and billions of gallons of gas — are wasted due to cars idling at stop lights. So, **IDLE AS LITTLE AS POSSIBLE**.

- If you think you will likely sit at a light for more than 30 seconds or so — such as at long traffic lights, withdrawing money at an ATM, waiting at a railroad crossing, etc. — consider turning the engine **OFF** (be prepared to buy a new starter more frequently, however). With a fuel-injected engine you'll use less gas starting the car again than is wasted while idling.

- If you often sit in drive-thru lanes, idling all the while, consider parking and going into the store to do your shopping, pick up your hamburger, etc. You'll save gas and your starter will thank you.

    Idling larger, more powerful engines (8- cylinder vs 4-cylinder, for instance) burns more gas; so folks with large SUVs, large trucks and high-horse-power engines, take note.

- When sitting at stoplights, railroad crossings, drive-thrus, etc. with the engine running, **shift your automatic transmission into NEUTRAL** rather than leaving it in **DRIVE**. This not only saves gas but allows transmission fluid to cool down (**be careful NOT to inadvertently shift into REVERSE when starting out again**).

When sitting at a traffic light, keep your foot off the accelerator completely. That way you'll be less likely to "rev" the engine while just sitting, waiting — you know, press on the accelerator just a little bit, usually unconsciously. "Revving" increases RPM and needlessly uses more gas. Keep your right foot on the **brake**!

Don't drive with your right foot on the accelerator and your left on the brake. Doing so, you're likely to "ride the brake" more often, wasting gas. Instead, **always use your right foot to operate *both* the accelerator and the brake**.

If it's not unsafe, turn **OFF** all auxiliary equipment you don't need to have **ON**, particularly when driving in town: headlights and driving lights, radio and disc players, GPS, etc. Each of these devices runs on electricity that's supplied by the alternator, and the harder the alternator works, the harder the engine works. More work = more gas used. Use power windows and mirrors, heater fans, and power seats sparingly.

CHANGE YOUR LIFESTYLE TO GET BETTER GAS MILEAGE

If you *really* **get serious** about improving fuel economy, do the following;

- Move out of the suburbs and nearer the urban core, where you will have commodities you need closer at hand, and where, when you need it, you can take advantage of whatever mass transportation is available.

- Live as close as possible to where you work, shop, play and where your kids go to school.

- Look carefully into what forms of **public mass transportation** are available in your area and use mass transit if it's convenient and matches your schedule.

- Consider taking your bicycle on public transportation to combine the advantages of each type of commuting. Buses in many communities now have bike racks installed on the front of each bus.

  For information on the public mass transportation alternatives available in your area, go to:

  - **American Public Transportation Association — http://www.apta.com**. This site provides information on all transit agencies and local links, by county & city.

- If suitable mass transit alternatives are scarce where you live, whenever possible leave the car in a "park-and-ride" lot and **carpool** or **ride-share**. If you and a buddy are riding together, the fuel cost of each of you is half what you'd each pay if driving your own cars, car maintenance expenses are reduced, and you only have the hassle of driving a few days a week. And the percentages get even better if you've got a group of three or four folks along for the ride.

  When you carpool, except on the days when you drive you don't have the hassle or expense of parking your car.

  Try to carpool to work at least once a week, and at work carpool with fellow workers when going out for lunch.

  To get hooked up with folks in your area who are looking to share rides to work (approximately the same destination), for shopping, car-sharing, and for longer trips (vacations, weekend commutes, overnight/weekend trips, etc.), visit the following websites:

  - Google "**RideShare Programs (fill in the city, metro area, etc)**" for a list of existing ride sharing opportunities in a designated locale.

- **http://www.eRideShare.com**. One of the most popular ride-sharing sites. Provides contact information of those wanting to share rides in a specific area (city, zip code, etc.).

- **http://www.AlterNetRides.com**. If you're interested in setting up any of a number of different types of ride-sharing programs in your area, this site provide the tools you need to organize and run it.

- Contact the **Metropolitan Transit Authority** in your city to get a list of existing carpools.

- Check on **community email listservs and bulletin boards** to get in contact with others with whom you share a common driving destination.

Another advantage to carpooling: with two or more people in your car you are qualified to legally use designated **High-Occupancy Vehicle (HOV) lanes**, where you'll likely encounter less traffic congestion, travel faster, and spend less time enroute (less time on the road means less gas burned).

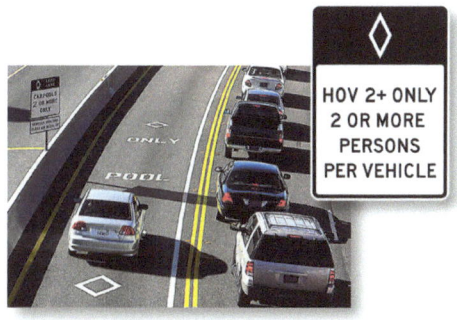

Clearly marked bike lane in Portland, Oregon.

- **Bikes use no gas at all**, so ride one when practical. Seek out and use routes to work, shopping, etc. where roadways have clearly marked bike lanes.

- Petition your city or county to create bike lanes where none now exist.

- Ask the company where you work to install bike racks so people feel safer about riding their bikes to work. You may even ask them to build showers and install lockers for folks who walk, jog or ride bikes to work.

In many cities you can strap your bike to the front of a municipal bus, making good use of two great gas-saving techniques.

Installing a rear rack and pantiers give you the ability to cart all your "stuff" on your bike.

- An alternative to a traditional bike is a **motorized version** that runs on either a small 48-80cc gasoline-powered engine, or electrical batteries. 2-stroke and 4-stroke models are available. Both can achieve speeds of 30-40 mph and get about 120 miles per gallon. They are fully street-legal for use in bike lanes.

Image of motorized bicycle courtesy of Helio Bicycles

- **Consider Riding a Segway Personal Transport Vehicle.** The gyro-stabilized Segway PT can take you places a car or bicycle can't — inside stores, office buildings, businesses, airports, elevators, and trains. Although ideal for short jaunts, Segway PTs can travel 24 miles on a single battery charge, depending on terrain, payload, and riding style.

Segway offers 7 different models, including those designed specifically for normal terrain, varied terrain, and a "commuter" model. The "cargo" model even adds saddlebags to the "commuter" model to carry more of your gear.

An all-electric vehicle, the zero-emissions Segway PT is 11 times more efficient than the average American car, and over three times more efficient than even the highest-mileage scooters.

- **Mopeds** have been popular in Europe for decades. Now may be the time to consider riding one around your town. Powered by 50-300 cc gas engines, they come in 2- and 3-wheel versions and are great for short urban trips. They can carry two people and get about 100 miles per gallon.

  Electric versions use batteries, and travel about 25 miles on a 20-cent charge.

- Consider **telecommuting** (working from home). Negotiate with your employer to allow you to work from home as often as practicable, if your job is well suited and your employer is amenable.

- **Walk** when the distance you must travel is less than a mile or two, and the weather is nice. Carry a day-pack for carting around groceries, shopping treasures, etc.

- **Pay your bills online** to avoid having to drive to the post office to mail your payments.

- If possible, **do your banking at a facility that's within walking distance of your home or office** so you can make deposits or use an ATM without having to drive to the bank.

- Have your paycheck and other regular deposits (annuity payments, pension checks, etc.) **direct-deposited** to your bank account, so you don't have to drive to the bank and sit idling in a drive-through line.

- As often as possible, use **email**, **telephones**, **faxes** and **other electronic communications** to do business with clients and communicate with your office and friends.

- Rather than traveling to meetings, **promote the use of teleconferencing software** when it's necessary for two or more people to share information and/or images on their computers. This includes web conferencing, online meetings of various sized groups, and desktop sharing and collaboration. Software that makes this possible includes, among others:

    Go To Meeting:
    http://www.gotomeeting.com/fec/

    Cisco WebEx:
    http://www.webex.com/

    iMeet:
    https://www.imeet.com/

    Adobe Connect:
    http://connect.brand.us.sem.adobe.com/contentry?sdid=IEASO&skwcid=TC|22191|adobe%20connect||S|e|5894715982

    Microsoft Office Live Meeting:
    http://www.livemeetingplace.com/livemeeting/?gclid=COr_x72g1KcCFQjd4AodMzLv9g

Yugma:
https://www.yugma.com/

and, at the high price end, **Cisco's** Telepresence *system:*
http://www.cisco.com/en/US/products/ps7060/index.html

- Do your shopping research — and shopping — **ONLINE**. These days you can buy nearly *anything* you desire on the Internet (from clothing, to shoes, to home electronics) and have it shipped directly to your front door — all without you burning a single drop of gasoline.

  Save even more gas by doing your pre-order research and product comparisons online or on the telephone. Reading reviews by folks who have purchased and used a product is a valuable advantage of online shopping, as well.

  When an item you purchase is to be delivered, be sure to select the "Standard" delivery option; it'll take longer to receive your item, but you'll be saving gas for the delivery company, and hopefully contributing to the reduction of carbon dioxide emissions, as well.

- Shop at the Mall and other locations where you can easily *walk* from store to store; you'll reduce the amount of gas you'll spend driving from shop to shop and looking for parking spaces.

- Rather than driving to your local video store to pick up a movie, consider signing up for a video service, such as Netflix or Blockbuster, that let's you select a movie and delivers it right to your mailbox within a couple of days. As a Netflix customer (about $10.75 a month) you can also watch thousands of free streaming movies on your TV or computer.

BUY A MORE FUEL EFFICIENT CAR OR ONE THAT RUNS ON ALTERNATIVE FUEL

This series of Tips has assumed that your present vehicle has a standard internal combustion engine, and that you will continue to drive it for at least a short time. When the time comes to purchase a different car, however, you may want to consider purchasing a more fuel efficient model or an Alternative Fuel Vehicle ("ATV"), one that operates on one of several alternative fuels.

Among the most fuel-efficient automobiles powered by gasoline-burning internal combustion engines currently on the market (2012) are the following:

Audi A3 (diesel)	Hyundai Elantra	Mini Cooper
Chevrolet Cruze Eco	Hyundai Sonata Hybrid	Nissan Altima Hybrid
Fiat 500	Kia Forte Eco	Smart4Two
Ford Escape Hybrid	Kia Rio	Toyota Camry Hybrid
Ford Fiesta SFE FWD	Lexus CT 200h	Toyota Prius Hybrid and V
Ford Fusion Hybrid FWD	Lexus HS 250h	Toyota Scion xD and iQ
Honda Civic Hybrid	Lincoln MKZ Hybrid	Toyota Yaris
Honda CR-Z	Mazda 2	Volkswagen Golf
Honda Fit	Mazda Tribute Hybrid	Volkswagen Golf (diesel)
Honda Insight Hybrid	Mercury Mariner Hybrid FWD	Volkswagen Jetta (diesel)
Hyundai Accent	Mercury Milan Hybrid FWD	
	Volkswagen Jetta Sportswagen	

If you're in the market for a new car, consider the fuel efficiencies offered by Hybrid vehicles such as the Toyota Prius. But be aware of the higher initial cost of the car, plus potentially higher maintenance costs.

Diesel engines can deliver better gas mileage than Hybrids on long distance trips. Diesel-powered vehicles are more popular than Hybrids in Europe.

Alternative fuels currently being developed or in operation include methanol, ethanol, compressed natural gas, liquefied petroleum gas, and electricity (more exotic alternatives are also being researched, such as hydrogen and solar-powered vehicles).

The use of such alternative forms of energy production in a variety of vehicles is generally intended to reduce harmful pollutants and exhaust emissions, such as carbon dioxide.

The Federal Trade Commission requires labels on all new AFVs that provide the user with general descriptive information about the alternative power source and the vehicle's estimated cruising range using that alternative fuel. Before purchasing an ATV, do some serious research to discover how many miles an AFV is expected to travel on a tankful or a standard supply of a particular fuel since, gallon for gallon, some won't travel as far as gasoline-powered vehicles. Then, choose an AFV that suits your needs and lifestyle.

Hybrid Electric vehicles, such as the Toyota Prius and the Ford Escape Hybrid, are yet another option. Hybrid Electric technology is becoming well-established, and hybrid-powered vehicles, which combine the benefits of gasoline engines and electric motors powered by on-board electric batteries, offer improved fuel economy and increased power.

Fully electrically-powered vehicles, such as the 2012 Chevy Volt and the 2012 Nissan Leaf, are powered totally by the energy stored in large on-board batteries, and produce no emissions from their operation. Unfortunately, the infrastructure for charging the batteries, such as well-distributed charging stations, does not yet exist, leaving owners of these vehicles on their own in finding ways of recharging their cars.

Solar-powered cars, and vehicles powered by a number of other "exotic" power sources, are in various stages of development and may become available sometime in the future.

Reality Check Publishing will soon publish a "Reality Check Guide" ebook covering alternative energy cars in depth. Visit our website, or email us your contact info and we'll let you know when it's available:

Web: **http://www.tipsforbettergasmileage.com**

Email: **sales@realitycheckpublishing.com**.

HYPERMILING TECHNIQUES TO AVOID

Hypermilers, folks who are addicted to wringing the absolute *maximum* mileage from every drop of gas in their tank, regularly use a variety of techniques to improve fuel efficiency. Many of their techniques, including "coasting" (see page 17), have been discussed in this book. A hypermiling technique we highly recommend you avoid, however, is "**DRAFTING**."

Drafting (or slipstreaming), a technique that's widely used in sportscar racing, bicycle racing and speed skating, involves positioning a car very close to the rear end of another vehicle while both cars are traveling at high speed, so that the drafting vehicle is essentially *tucked inside the lead vehicle's slipstream*. The intent is to reduce the overall negative effect of aerodynamic drag on the car that's drafting by exploiting the lead vehicle's slipstream.

As the lead vehicle moves forward, it pushes aside the air flowing over it, creating turbulence in the flowing air which at high speed creates a **partial vacuum** behind the vehicle. This partial vacuum literally "pulls" the drafting vehicle f*orward*, resulting in the drafting vehicle using less fuel to cover the same distance compared to the lead vehicle (interestingly, the technique can also result in the lead vehicle consuming slightly less gas, as well, because the drafting vehicle slightly reduces the negative "suction" effect of the low-pressure region on the lead vehicle).

The *shape* of a vehicle determines how effective drafting is: aerodynamically-shaped vehicles with tapered rear ends, such as many of today's passenger cars, create little turbulence and therefore a very slight vacuum. Large, bulkily-shaped vehicles with square rear ends and little aerodynamic shape, such as trucks and eighteen-wheelers, push aside much more air and create a great deal of turbulence. That's why hypermilers are often seen drafting behind large transport trucks traveling at 70-75 mph on an Interstate highway — by doing so they significantly improve their gas mileage!

Although you can definitely save gas by drafting, in the real world of highway driving the technique is extremely dangerous: because the drafting vehicle is traveling at high speed in very close proximity to the lead vehicle, and the driver's view of the road ahead is effectively blocked by the bulk of the large truck, the drafting driver has very little time to react effectively should the lead vehicle find it necessary to brake suddenly or take evasive action to avoid a potential hazard ahead. In fact, the dangers involved in drafting are so great that hypermiling has come under fire from law enforcement and others due to claims of dangerous or unlawful behavior by some hypermilers, such as tailgating larger vehicles on freeways.

In addition to the danger involved, drafting can be very hard on your car. Since the drafting vehicle is traveling so close to the lead vehicle, road dirt is thrown into it by the lead vehicle's tires. The front end quickly becomes very dirty. In addition, gravel and other debris on the roadway can easily be picked up by the lead vehicle's tires and propelled into the front of the drafting car. Damage to windshields, chipping of paint, and denting of hoods from flying road debris is very common.

— WARNING —
Due to the inherent dangers associated with drafting, we definitely recommend that you do *not* use this technique as you attempt to improve your gas mileage!

THE 15 MOST FUEL-EFFICIENT CARS — 2011/2012

To get the most out of every gallon of gas, resolve to remain aware of new forms of transportation and vehicle fuels as they become available, and to, when possible, select the most energy efficient form as your primary mode of transport.

Below is a list of the 15 most fuel-efficient cars, as of mid-2012. Note that the majority—the ten models that offer the very best current fuel efficiency—are "Hybrid Electric Vehicles" (HEV), which means they are powered by a powerful electric powertrain in combination with a much smaller than normal internal-combustion engine that takes over in specific situations.

RATING	AUTOMOBILE NAME	MILEAGE CITY/HIGHWAY	COMBINED EPA MILEAGE
1	2012 Toyota Prius C Hybrid	City 53/Hwy 46	50 mpg
2	2012 Toyota Prius Hybrid	City 51/Hwy 48	50 mpg
3	2012 Honda Civic Hybrid	City 44/Hwy 44	44 mpg
4	2012 Toyota Prius V Hybrid	City 44/Hwy 40	42 mpg
5	2012 Lexus CT 200h Hybrid	City 43/Hwy 40	42 mpg
6	2012 Honda Insight Hybrid	City 41/Hwy 41	42 mpg
7	2012 Toyota Camry LE Hybrid	City 43/Hwy 39	41 mpg
8	2012 Toyota Camry Hybrid	City 40/Hwy 38	40 mpg
9	2012 Lincoln MKZ Hybrid	City 41/Hwy 36	39 mpg
10	2012 For Fusion Hybrid	City 41/Hwy 36	39 mpg
11	2012 Scion iQ	City 36/Hwy 37	37 mpg
12	2012 Chevrolet Volt	City 36/Hwy 37	37 mpg
13	2011 Lexus HS 250h Hybrid	City 36/Hwy 37	35 mpg
14	2011 Honda CR-Z	City 31/Hwy 35	33 mpg
15	2011 Hyundai Elantra	City 31/Hwy 35	33 mpg

FUEL ECONOMY LINKS

Automotive and auto fuel technologies are constantly, and rapidly, changing. Each year, new cars incorporate a wide variety of new systems, and many offer improved fuel economy. It's important, therefore, that drivers who are interested in getting the very best gas mileage make an effort to remain informed about changes taking place in the automotive marketplace.

The following links are a good place to start to keep abreast of how emerging technologies are changing the way we drive, and are making possible much better fuel economy than we've enjoyed in the past:

www.Fuel Economy.gov
http://www.fueleconomy.gov

Environmental Protection Agency — Fuel Economy
http://www.epa.gov/fueleconomy/

Aftermarket Retrofit Device Evaluation Program
http://www.epa.gov/otaq/consumer/reports.htm

Federal Trade Commission: FTC Consumer Alert
http://www.ftc.gov/bcp/edu/pubs/consumer/alerts/alt064.shtm

**Natural Resources Canada: Personal Transportation –
ecoEnergy for Personal Vehicles**
http://oee.nrcan.gc.ca/transportation/personal-vehicles-initiative.cfm

Transportation Research Board
http://www.trb.org

The Car Care Council
http://www.carcare.org

American Public Transportation Association
http://www.apta.com

The Tire Rack
http://www.tirerack.com

ScanGauge II and ScanGauge E, Automotive Fuel Economy Meter
http://www.scangauge.com/

MSD Ignition "DashHawk", Automotive Fuel Economy Meters
http://www.dashhawk.com/

PLX Devices "Kiwi", Automotive Fuel Economy Meters
http://www.plxdevices.com/